Introduction

AF735235

The spectacular shapes of Uluṟu (Ayers Rock) and Kata Tjuṯa (The Olgas) dominate the surrounding flat desert and are the culmination of geological events stretching over hundreds of millions of years. It is the description of those events — the telling of the geological story — that we attempt in this booklet. It may seem a complex and at times bizarre story, but a fascinating one nevertheless.

Archaeological research suggests that there has been human settlement in this region from at least 22 000 years ago, although the initial settlement was evidently much earlier. Reports from the first European explorers show that Aboriginal people regularly visited Uluṟu and Kata Tjuṯa. Both sites remain important base camps for the Pitjantjatjara and Yankuntjatjara people. Many features of Uluṟu and Kata Tjuṯa are an integral part of the ancient Aboriginal culture and some are so sacred that only Aboriginal people can visit them. Tjukurpa is the Aboriginal word describing the laws which give meaning and order to all aspects of life. The Tjukurpa provides explanations for the origins of life and all living things as well as features of the landscape. The science of geology is similarly based on sets of rules, but the interpretation of the origin of the landscape and its features are very different from those of the Tjukurpa.

Inalienable freehold title to Uluṟu (Ayers Rock - Mount Olga) National Park was handed back to the traditional owners in 1985 and is held by the Uluṟu Kata Tjuṯa Land Trust. The Trust has leased the Park to the Director of the Australian National Parks & Wildlife Service for 99 years for its continued use as a national park. With the hand-back, an agreement for joint management of the Park was arrived at. A commitment to use Aboriginal (Aṉangu) names for features or localities, rather than European ones, was a part of this agreement.

Kata Tjuṯa

Colin Totterdell

The first geological observations: European explorers, Giles and Gosse

The first sighting of Mount Olga by a European was on 13 October 1872, when Ernest Giles, exploring the country some 100 km to the northeast, climbed a ridge and looked to the southwest. 'After a long and anxious scrutiny through the smoke and haze, far, very far away, a little to the west of south, I descried the outline of a range of hills, and right in the smoke of one fire an exceedingly high and abruptly ending mountain loomed.' Giles's progress towards the mountain was barred by Lake Amadeus. Later he named Lake Amadeus and Mount Olga after the then King and Queen of Spain.

The following year Giles reached Mount Olga in a second expedition, but was bitterly disappointed when, at the foot of the mountain, he came upon dray, horse and camel tracks, those of explorer W. C. Gosse, who had reached the area some two months before him. Gosse, who had also approached the region from the north, skirted the eastern end of Lake Amadeus, and on 18 July 1873 wrote '... I could see ... Mount Olga, and a hill east of it, also a high flat-topped hill which I have named Mount Conner ...'. Clearly he had not realised at that stage how spectacular a sight Uluṟu is, for on the following day, as he approached he wrote 'The hill as I approached, presented a most peculiar appearance, the upper portion being covered with holes or caves. When I got clear of the sandhills, and was only two miles distant, and the hill for the first time, coming fairly into view, what was my astonishment to find it was one immense rock rising abruptly from the plain; the holes I had noticed were caused by the water in some places forming immense caves.' He named it Ayers Rock, after Sir Henry Ayers, the then Premier of South Australia.

Mount Olga is the more wonderful and grotesque, Mount Ayers the more ancient and sublime.'

Gosse climbed the rock and wrote 'This is a high mass of granite, the surface of which has been honeycombed, and is decomposing ...' When Giles visited Uluṟu nearly a year later, he wrote: 'The great difference between it and Mount Olga is in the rock formation, for this is one solid granite stone, and is part and parcel of the original rock, which having been formed after its state of fusion in the beginning, has there remained, while Mount Olga has been thrown up subsequently from below. Mount Olga is the more wonderful and grotesque, Mount Ayers the more ancient and sublime.' In fact, Uluṟu is not composed of granite but of a type of sandstone called arkose.

Giles's initial description of Kata Tjuṯa was accurate. On 14 September 1873 he wrote: 'The appearance of this mountain is marvellous in the extreme, and almost baffles description. ... it is formed of several vast and solid rounded blocks of bare red-coloured stone, of a kind of conglomerate, being composed of untold masses of round stones of all kinds, mixed as plums in a pudding, and set in vast and rounded shapes upon the ground.'

Giles gave free rein to his imagination when he returned to Kata Tjuṯa in June the following year, after an heroic attempt to find a passage through the desert to the west. He speculated that 'it is, I should say, decidedly of volcanic origin, belched out of the bowels and on to the surface of the earth by the upheavings of subterraneous and subaqueous fires, and cooled and solidified into monstrous masses by the gelid currents of the deepmost waves of the most ancient of former oceans.' An original and imaginative hypothesis indeed, but one which is not borne out by the facts!

Gosse, who reached The Olgas on 8 August 1873, wrote, on reaching the first outcrops, 'This range is formed of a number of round-topped masses of solid conglomerate rock (known as pudding-stone)... Each hill is a separate rock.' He scrambled 900 feet (270 m) up the steep slopes of Mount Olga, and 'could have reached the top, but I was not so sure about being able to get down again.' Ernest Giles wrote of Mount Olga in a similar vein and on 14 September 1873 recorded: 'I can only liken it to several enormous rotund or rather elliptical shapes of rouge-mange, which had been placed beside one another by extraordinary freaks or convulsions of nature.'

Giles was not a geologist and thus resorted to his fertile imagination to explain these splendid geological features.

Ernest Giles: 'Mount Olga from 60 miles to the west'

National Library of Australia

Ayers Rock (Uluṟu), sketched by William Gosse

Mortlock Library, South Australia

Part of William Gosse's map showing his exploration route in central Australia

National Library of Australia

Modern geological studies

Systematic documentation of the geology of the region began in the early 1900s, and culminated in the geological mapping of the whole of the southern Northern Territory by the Bureau of Mineral Resources, Geology & Geophysics (BMR) between 1960 and 1964. The area encompassing Uluṟu and Kata Tjuṯa was mapped in 1963. The major rock outcrops were examined in that survey, and provided the basic framework of geological facts on which many of our interpretations are based. More recently the examination of rock chips from water-bore drilling, and closer examination of the rocks at Uluṟu and Kata Tjuṯa, have led to new insights into the way in which the rocks were deposited. Studies of variations in the earth's magnetic field have also yielded valuable information (see below).

It is important to realise that the geological history of the region is interpreted from many sets of incomplete data — rather like trying to draw a picture from a jigsaw puzzle in which many (perhaps even most) of the pieces are missing. We believe that the 'picture' we present is substantially correct, but there is no doubt that it will be modified as further mapping is done, as more water-bore and geophysical information becomes available, and as new concepts in geology are applied.

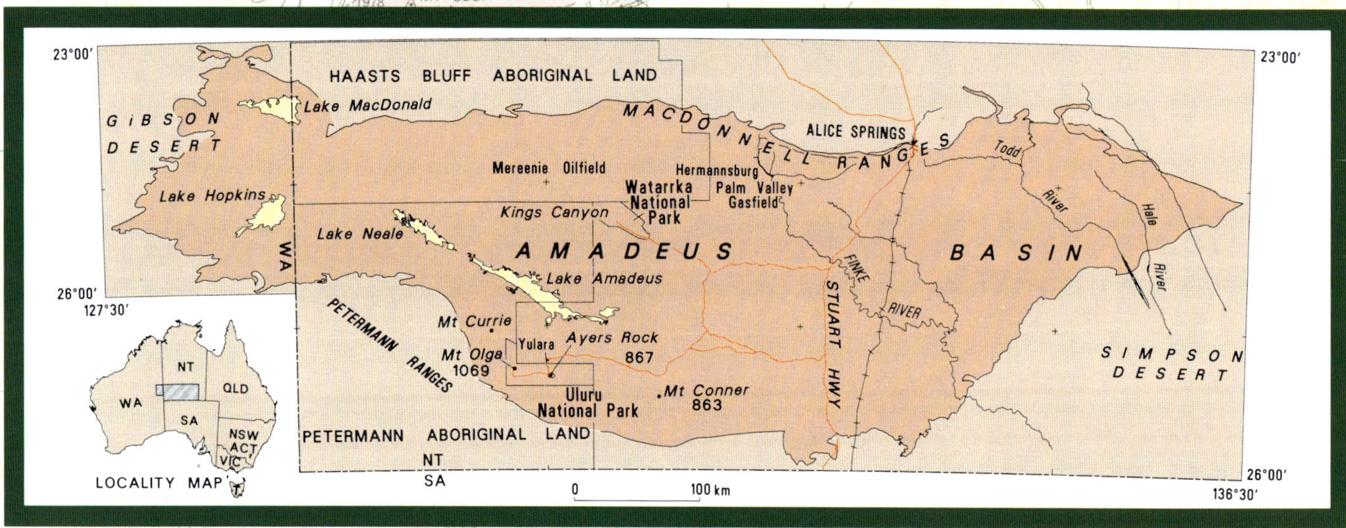

Uluṟu and Kata Tjuṯa lie near the southwestern margin of the Amadeus Basin, in central Australia

Remote sensing and geophysics

Remote sensing refers to techniques of measuring various properties of the earth's surface from a distance. Perhaps the best known techniques is that of satellite imagery, in which various bands of electromagnetic radiation either reflected or emanating from the earth's surface are measured, and used to produce images. Such imagery can be very useful for outlining major geological features of the earth's crust

Geophysical techniques are very powerful tools, in many cases the only tools, for studying the interior of the earth. Common techniques include measurement of the strength of the earth's gravity field, usually carried out on the ground, and radiometric and magnetic surveys which are commonly carried out from low-flying aircraft with appropriate sensors on board. The aeromagnetic technique has proved useful in the Uluṟu region, as it can detect magnetic rocks beneath the thin sand blanket and give us some idea of the bedrock geology.

Geological History

The beginning — the formation of the Amadeus Basin

Between about 900 million and 600 million years ago, much of central Australia lay at or below sea-level, forming a depression, an arm of the sea, known as the Amadeus Basin. Rivers carried mud, sand and gravel into the depression, building up layers of sediment. The sediment did not reach all parts of the depression, and in those areas that were starved of sediment, chemical deposits, later to become limestone and dolostone (see glossary, p 26) accumulated instead. At times large areas of the basin became at least partly blocked off from the sea, resulting in evaporation of the water in it. Layers of salt were formed, and were eventually covered by limestone and dolostone, and by more sandstone and mudstone. Evidence of very cold climates is also preserved, as some of the sediments were deposited by glaciers. Glacial rocks of similar age are also preserved in the Flinders and Mount Lofty Ranges in South Australia, and in the Kimberley region in Western Australia, suggesting that a major ice age affected the Australian region about 700 million years ago.

The floor of the Amadeus Basin, and much of the earth's crust beneath it, must have sagged downwards at about the same rate as the sediments built up, because much of the 3 to 4 km of sediment that accumulated in the basin was apparently deposited in very shallow water. As the sediments were buried, trapped water was squeezed out, and the strata were compressed and cemented to form sedimentary rocks. Sand became sandstone, mud became mudstone or shale, and salt became rock salt, and so on. The eroded remnants of many of these sedimentary rock formations are visible throughout central Australia today: sandstone forms the spectacular ridges of the Macdonnell Ranges east and west of Alice Springs and Mount Conner east of Uluru, whereas limestone, dolostone and shale form many of the lower hills. Rock salt is not present in outcrops because it is readily dissolved in groundwater.

900 to 600 million years ago - areas of central Australia lay under water

Geological time scale

Eon	Era	Period	Epoch	Duration	Age*
PHANEROZOIC	CAINOZOIC	Quaternary	Recent	0.01	0.01
			Pleistocene	1.8	1.8
		Tertiary	Pliocene	3.5	5.3
			Miocene	18.2	23.5
			Oligocene	13.2	36.7
			Eocene	21.3	58.0
			Paleocene	8.4	66.4
	MESOZOIC	Cretaceous		77.6	144
		Jurassic		69	213
		Triassic		35	248
	PALAEOZOIC	Permian		38	286
		Carboniferous		68	354
		Devonian		54	408
		Silurian		26	434
		Ordovician		71	505
		Cambrian		85	590
PROTEROZOIC	NEOPROTEROZOIC			410	1000
	MESOPROTEROZOIC			600	1600
	PALAEOPROTEROZOIC			900	2500
ARCH-AEAN			(greatest age so far measured)		4276
				(age of the earth)	4500

* Age of base of era, period or epoch in millions of years

Sand dunes begin to form

Start of present arid cycle

Deposition of sediments in valley between Uluru and the Kata Tjuta basins

Alice Springs Orogeny

Deposition of sands and boulders in alluvial fans forming the Uluru Arkose and Mt Currie Conglomerate

Petermann Ranges Orogeny

Amadeus Basin forms

Time Span of Amadeus Basin

Origin of the strata at Uluṟu and Kata Tjuṯa

About 550 million years ago, in the Cambrian Period, the southwestern margin of the Amadeus Basin was raised above sea-level, ending the widespread sedimentation in that area. The rocks were squeezed, crumpled and buckled into folds ranging from millimetres to kilometres across, and fractured along faults. This episode is called the Petermann Ranges Orogeny, and it affected all of the region from Uluṟu west and south into Western Australia and South Australia.

550 million years ago - the rocks were squeezed, crumpled and buckled into folds

During the later stages of the Petermann Ranges Orogeny the country was uplifted and subjected to very rapid erosion to form a mountain range. The sedimentary rocks of Uluṟu and Kata Tjuṯa are the products of this rapid erosion, and they represent a new phase of development of the Amadeus Basin. Although the rocks at each place look different, in each case they are typical alluvial fan deposits. They are unlike most sandstones laid down on beaches or in the sea in that they are poorly sorted (i.e. there is a mixture of grains of vastly different size), and the structures within the rocks suggest that the sediment was deposited rapidly. The difference in character between Uluṟu and Kata Tjuṯa is best explained by deposition on different alluvial fans.

Alluvial fans built up during the Petermann Ranges Orogeny, some 550 million years ago

The alluvial fans were very thick. At Uluṟu, the layers total at least 2.5 km in thickness, indicating that the mountain range from which the sediment came must have been gradually uplifted by at least that much. However, the mountains may never have actually towered 2.5 km above the adjacent alluvial fans. The fans would have started to build up as soon as mountains began to rise, and would have continued to build up and out from the mountains. At the same time the area over which the fans were being deposited was subsiding, perhaps at only a few centimetres

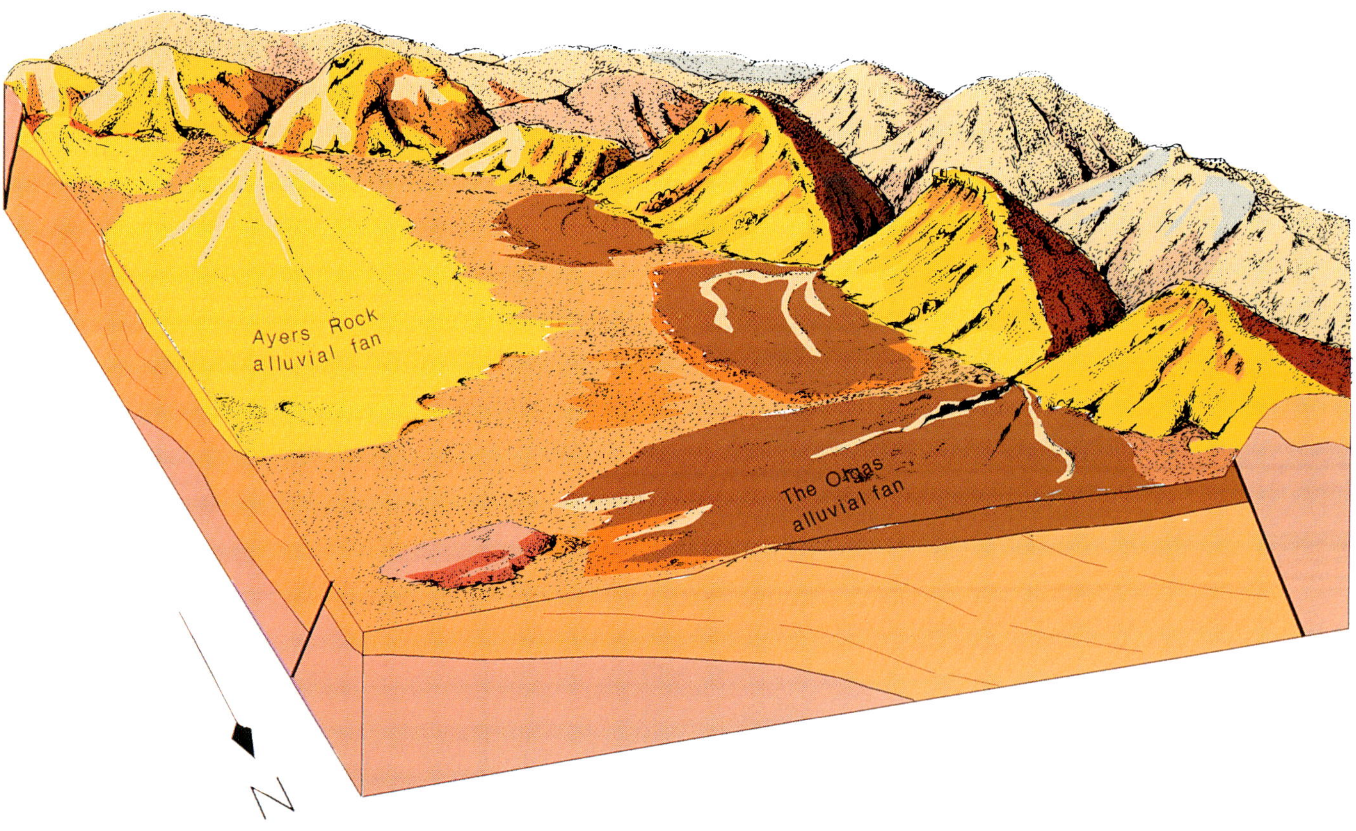

Colin Totterdell

Clast imbrication - the slope of the boulders at the centre indicates the direction from which the conglomerate was derived

a year, but, over the thousands of years that deposition of the fans took place, a large thickness of sediments accumulated. The conglomerate at Kata Tjuṯa is even thicker — perhaps as much as 6 km. Its thickness and coarseness suggests that it represents part of a fan much closer to the ranges. The ranges may have been higher at this point and the streams that carried such large boulders (some are over 1.5 m across) were very swiftly flowing. Geologists refer to the conglomerate formation at Kata Tjuṯa as the Mount Currie Conglomerate (it is also well exposed at Mount Currie, 35 km to the northwest).

The 'Uluṟu Arkose alluvial fan' at Uluṟu was a plain that sloped gently to the north for many tens of kilometres. Instead of being cut by large creeks or rivers, it rather consisted of a vast network of very shallow sandy channels, all less than 1 m deep. Heavy rain in the mountains caused flash floods, which, heavily laden with sediment, rushed down deep gorges and out onto the plain. Here, the absence of valley walls caused the water to spread out as 'sheet floods', each carrying huge volumes of sand onto the plain, spreading it out for many kilometres as sheets some centimetres thick. The surface of the sand was sculpted into ripples such as those preserved in the intricate lamination seen at Malaku Wilytja.

The 'Mount Currie Conglomerate alluvial fan' at Kata Tjuṯa may have been even larger and more spectacular than the one which formed the Uluṟu Arkose (at Uluṟu), but most of it has been eroded away, leaving only those parts which originally flanked the ancestral Petermann Ranges.

Colin Totterdell

Cross-bedding at Malaku Wilytja

Huge volumes of sand and gravel spread across the alluvial fan surfaces

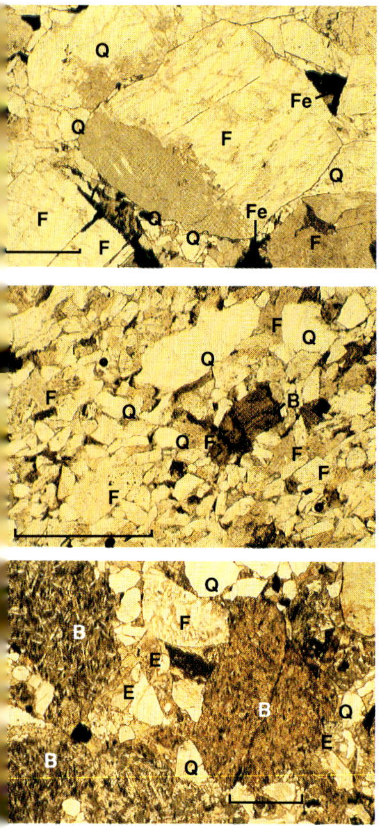

Uluṟu Arkose and Mount Currie Conglomerate under the microscope: (a) large sand grains of feldspar (F) and quartz (Q) set in a matrix of small quartz grains and iron oxides(Fe). (b) fine-grained, well-sorted arkose. (c) Poorly sorted sandstone matrix from the Mount Currie Conglomerate; speckled dark grains are basalt(B), separated by quartz and feldspar grains and irregular patches of epidote (E); the scale bar is 1 mm long

The rocks themselves

The rock making up Uluṟu is an arkose. It consists mainly of sand and small pebbles of quartz and feldspar, with minor amounts of clay, iron oxides, and tiny fragments of other rocks. It is typical of a sediment that has been rapidly eroded from mountains made of granite or similar rock. In fact, the arkose looks remarkably like granite, even though it is a sedimentary rock.

Kata Tjuṯa consists of conglomerate, composed mainly of very large 'clasts' — pebbles, cobbles and boulders that are recognisable as pieces of rock, rather than being single mineral grains. The main types of rock are granite and basalt, but there are also a few boulders of sandstone and rhyolite in the lower conglomerate beds, and several kinds of metamorphic rocks throughout.

The Cambrian landscape

The sandstone and conglomerate forming Uluṟu and Kata Tjuṯa were laid down at a time when many of the plant and animal groups with which we are familiar today did not exist — they had not yet evolved. At the beginning of the Cambrian Period, some 590 million years ago, the sea was inhabited only by animals without backbones (invertebrates) — sea snails (gastropods), lamp shells (brachiopods), sponges, trilobites, various kinds of worms and primitive sponge-like organisms that looked like coral (archaeocyathids). There were no fish: they were to evolve about 100 million years later, during the Ordovician.

On land the scene was just as dramatically different: there were no plants as we know them, i.e. vascular land plants, which includes all trees and grasses. The landscape must have been devoid of the green hues with which we are so familiar. However, bacteria and algae had evolved hundreds of millions of years earlier, and would have dominated any hospitable niche, such as moist clay or sand, cracks in rock surfaces, and lakes or ponds. They must have contributed to the weathering of rocks, and would have hastened the erosion of mountainous regions, but the absence of trees would have meant that there was little organic matter to rot down and form soil. Bare rocky ranges were common, even in regions of high rainfall; erosion was rapid because the lack of vegetation enabled water to drain off the slopes immediately, resulting in flash floods carrying gravel, sand, and clay to adjoining lowlands.

The alluvial fans are buried

The alluvial fans were buried beneath sand, mud and lime rich sediments

As the ancestral Petermann Ranges were eroded, building of the alluvial fans diminished. By about 500 million years ago the region was again covered by a shallow sea and the alluvial fans were gradually buried beneath sand, mud and limestone. North of Uluṟu, huge volumes of sand were laid down along the shoreline of this sea, and the sand was later buried by mud. Many kinds of animals lived in the sea, and as they died, their remains settled to the sea floor and were gradually buried by sediment. One particular mud-rich sedimentary layer, the Horn Valley Siltstone, contains a great number and variety of fossils; this siltstone is a rich petroleum source, and the oil and gas which formed in it accumulated in the underlying Pacoota Sandstone. The Pacoota Sandstone contains the Mereenie Oil Field and Palm Valley Gas Field, southwest of Alice Springs, which supply much of the Northern Territory's energy needs.

By Silurian time, about 430 million years ago, the sea had completely receded from the Amadeus Basin, and wind-blown sand began to accumulate. That sand is now the Mereenie Sandstone, which is very porous and capable of storing large volumes of water. It is the main source of groundwater for Alice Springs. The Mereenie Sandstone also forms the walls of Kings Canyon. Overlying it are siltstone and mudstone deposited during the Devonian Period, some 390 million years ago.

The Alice Springs Orogeny

A new period of folding, faulting and uplift, more widespread than the Petermann Ranges Orogeny, began around 400 million years ago, and continued, with some pauses, for around 100 million years. In this 'Alice Springs Orogeny' the layers of conglomerate and arkose at Uluṟu and

400 million years ago - a new period of folding, faulting and uplift

Kata Tjuta, which had been buried by hundreds or even thousands of metres of younger Amadeus Basin sediments, were strongly folded and faulted, and the originally horizontal layers at Uluru were rotated into a nearly vertical position.

The Mount Currie Conglomerate at Kata Tjuta was less affected; it dips to the southwest at only about 15 degrees. It is faulted along its southern edge, and the dip changes near the fault, showing that it was dragged upwards along the fault.

The Alice Springs Orogeny also tilted, folded and faulted the rocks which form the Macdonnell Ranges.

The final stages

Uplift during the Alice Springs Orogeny raised much of central Australia above sea-level and a new period of erosion commenced. The area that is now Uluru (Ayers Rock - Mount Olga) National Park has probably remained above sea-level since that time — for some 300 million years.

Erosion has continued for over 300 million years

Initially the land surface was much higher than the top of Uluru and Kata Tjuta. As erosion continued, Uluru, Kata Tjuta and Mount Conner formed part of a watershed between Lake Amadeus to the north and the country to the south. About 70 million years ago, a broad shallow valley had been eroded between Uluru and Kata Tjuta. Uluru and Kata Tjuta were present as high ground flanking the valley, but whether they appeared similar to their present shape at that time is a matter of conjecture. The valley gradually filled with sediments deposited by streams and in marshes. Plant material accumulated in the marshes to form layers of peat, which developed into thin layers of lignite (brown coal) as they were compressed. The marshes and streams were part of an ancient river system that drained east towards the present-day Finke River.

Some of the brown coal contains fossil pollen and spores from land plants which are found today growing in wetter conditions. The oldest brown coals are 65-70 million years old, and contain an abundance of the distinctive, two-winged pollen grains of southern conifers, *Podocarpus*, which suggests that rainfall was high. Fern spores are common too. There is also pollen of a plant known as *Gunnera*. Modern representatives of *Gunnera* are giant herbs with prickly rhubarb-like leaves; all grow today in very wet climates, mainly in rainforests in New Guinea and central and southern South America.

Beech forests in the heart of Australia suggest moister climates than now

The youngest pollen grains found in these sediments are about 40 million years old. Pollen suites of this age are similar to others found in the Alice Springs area, and contain abundant pollen of *Nothofagus*, the southern beech or myrtle beech, which grows now in Tasmania and coastal eastern Australia. The presence of beech forests in the heart of Australia at this time suggests much moister climates then. Other pollen types include extinct members of the Proteaceae, the very large family that includes the banksias and grevilleas. There are also rainforest species belonging to the family Myrtaceae. Today this family is best known for the eucalypts, but these were not in evidence at this time; only their rainforest relatives were present in central Australia. Gum trees came to dominate the Australian landscape much later, possibly as recently as 2 million years ago.

Sand layers in these younger sediments are the main aquifers (water-bearing strata) in this area and today bores drilled into them provide the water for Yulara. The water is slightly salty, and has to be desalinated before use. The water itself has been dated using the carbon-14 technique and its age ranges from modern to 7000 years. The older water comes from several periods when the climate was wetter. However, most of the groundwater is modern and derived from recent rainfall.

The water table is 25 m deep near Kata Tjuta but shallows to 12 m at Yulara airport. The groundwater eventually flows underground into Lake Amadeus.

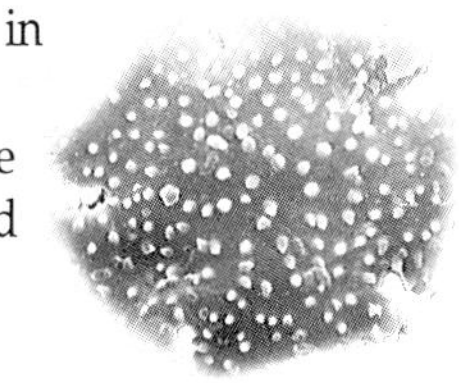

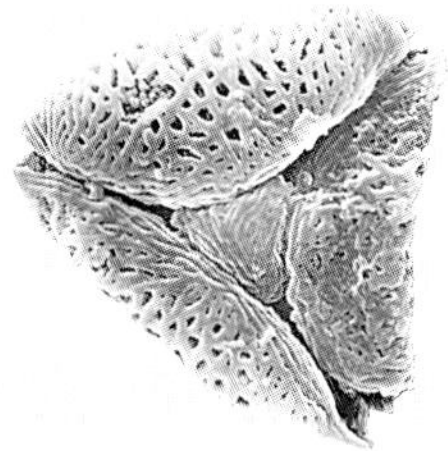

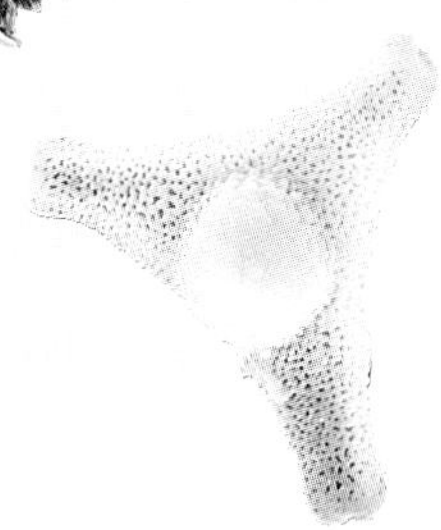

Fossil pollens magnified one thousand times

About 500 000 years ago the climate became more arid, and this has continued to the present day, although interrupted by several slightly wetter episodes. Wind-produced deposits form either gently sloping plains with occasional calcrete outcrops, or systems of irregular, reticulate or parallel dunes with stable flanks and enclosed swales containing alluvium, calcrete and gypsite. The sand dunes seen in the Park today are up to 13 m high, and dating by the thermoluminescence technique has shown that they began forming about 30 000 years ago.

Colin Totterdell

Sand ridges, partly stabilised by vegetation, cover vast areas in inland Australia

Bedrock geology

The sketch map is a preliminary interpretation of the bedrock geology, incorporating information from outcrops, water-bore drilling, and aeromagnetic contour maps. It shows the rocks that we would expect to see if we could strip the sand and older Cainozoic sediments from the region. It shows in the southwest a broad band of highly magnetic rocks comprising 'basement' (the ancient igneous and metamorphic rocks which form the floor of the Amadeus Basin), and the older Proterozoic rocks of the Amadeus Basin. Three small areas of these magnetic rocks are present near major faults between Uluṟu and Kata Tjuṯa. Magnetic rocks present at shallow depths northeast of Yulara may be thin basalt layers within older parts of the Proterozoic sequence of the Amadeus Basin.

Overlying the older rocks are the younger Proterozoic rocks of the Amadeus Basin; they are non-magnetic. A broad band of these is interpreted south and southwest of Kata Tjuṯa. Parts of the Mount Currie Conglomerate (which forms Kata Tjuṯa) are slightly magnetic, perhaps due to the presence of the mineral magnetite in the conglomerate, and a very broad band of the conglomerate is confidently interpreted to extend from Fault A through to Mount Currie in the northwest. The Uluṟu Arkose is not magnetic, and its extension to the northwest beneath the sand is based on limited outcrop and water-bore information. There is some evidence that other bands of it are present to the west, as shown in the cross-sections. The thrust fault, F, is inferred from the presence of very magnetic rocks to the southwest, similar to the basement. Fault A may be an old fault related to the Petermann Ranges Orogeny (see below), or it may be a thrust fault generated later during the Alice Springs Orogeny. Faults B to F developed during the Alice Springs Orogeny.

This sketch map, or 'solid geology' map, is an excellent tool for determining where further studies could assist our understanding of the geology of the region. For instance, valuable information would be gained if future water bores north of the Rangers Station at Uluṟu, or southwest of Yulara, were drilled deep enough to penetrate bedrock. The lines marked K-K' and U-U' show the positions of the series of cross-sections described below.

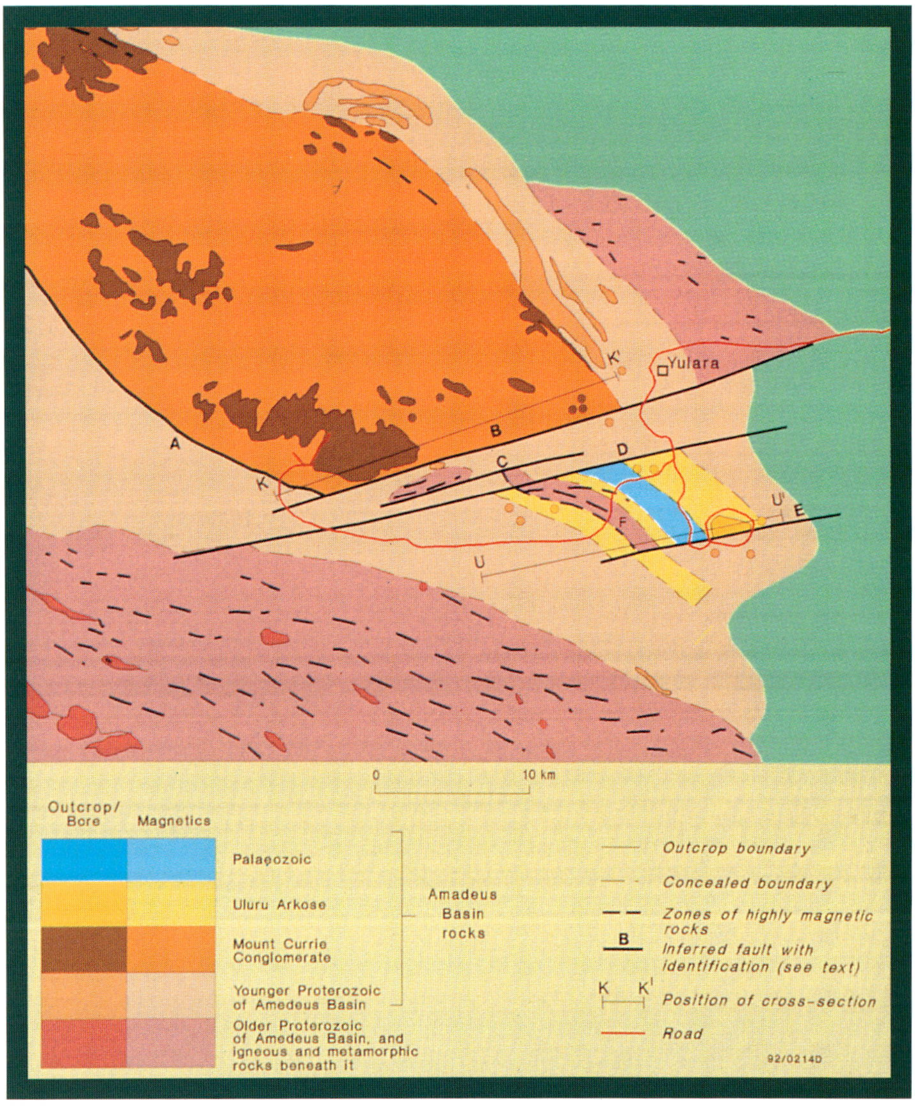

Bedrock geology map, an interpretation of rocks beneath the sand cover

Pictorial summary of the geological history

The following diagrams are cross-sections through the earth's crust down to a depth of several kilometres. The sections are composite, each consisting of two separate sections — one trending east-northeast through Kata Tjuṯa, the other trending in the same direction through Uluṟu. The positions of the sections are shown on the bedrock geology map. The rectangles show the future positions of Uluṟu and Kata Tjuṯa.

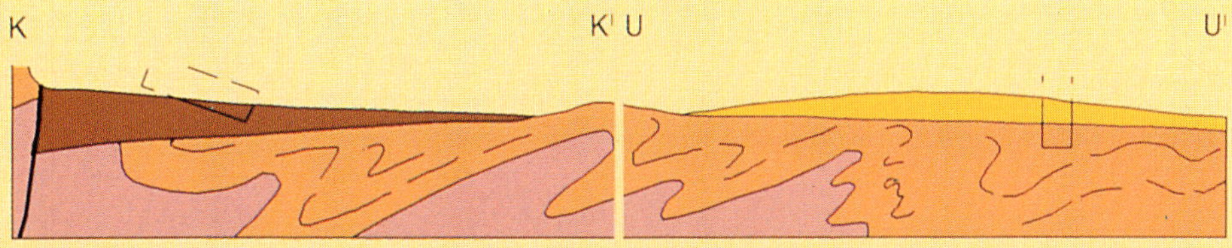

During the final stages of the Petermann Ranges Orogeny, (550 million years ago) alluvial fans build out from ancient mountains, forming extensive wedge-shaped bodies of conglomerate (dark brown) and arkose (yellow), on a "basement" of folded Proterozoic sedimentary rocks (brown) and igneous and metamorphic rocks (pink)

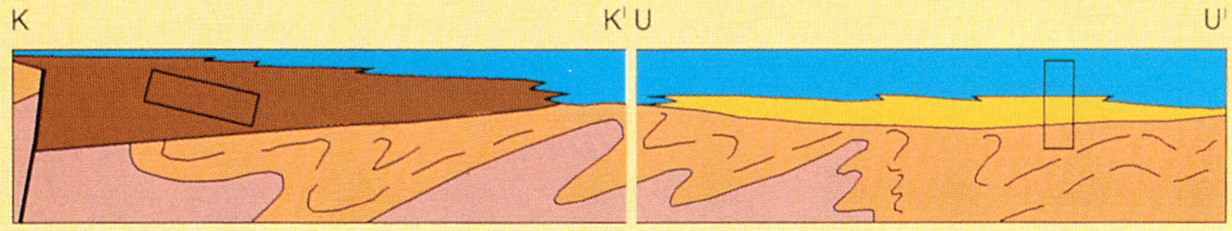

Fan deposition wanes during the early Palaeozoic as the mountains are eroded. The region sinks, and as the sea encroaches a thick sequence of sand, mud and limestone (blue) is deposited.

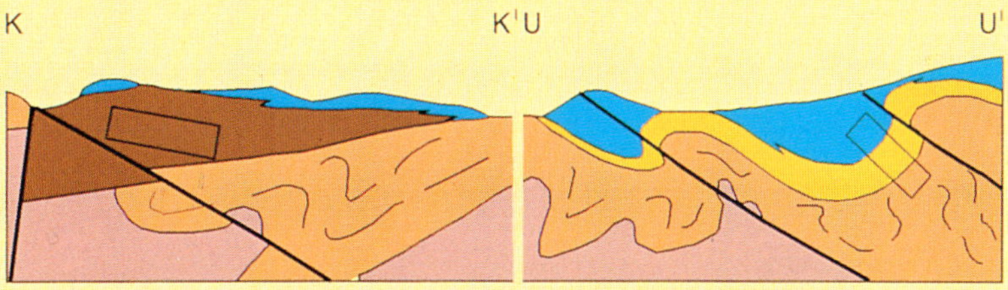

During the mid-Palaeozoic major faults and folds begin to develop in the rocks (the Alice Springs Orogeny, 400-300 million years ago). The layers of Mount Currie Conglomerate and Uluru Arkose begin to tilt and fold into their present positions, under the influence of huge "thrust" faults. The deformation results in considerable sqeezing/shortening of the region due to enormous compressive forces. Rapid erosion occurs during deformation of the rocks.

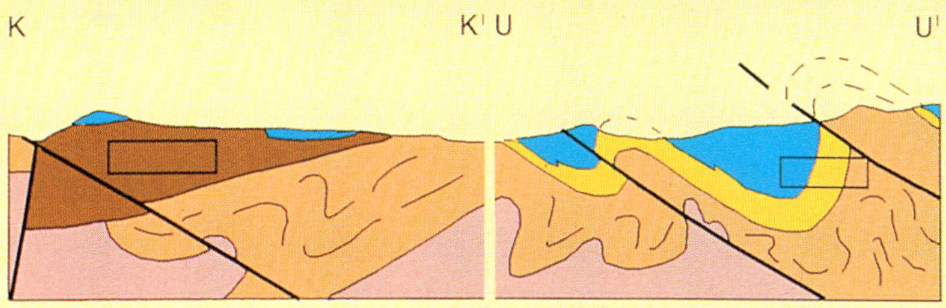

After the peak of the Alice Springs Orogeny, in the late Palaeozoic, folding and faulting wane, but erosion continues at a gradually reducing level.

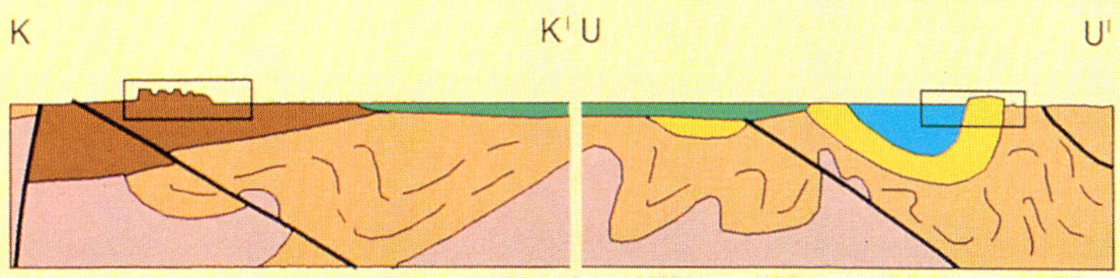

By about 65-70 million years ago a broad shallow valley had formed between the larger, but already shaped, Olgas and Ayers Rock. Layers of alluvial sediments (green) containing some thin brown coals were laid down. By 30 000 years ago sand dunes, seen today, had formed.

92/02 15D

The shaping of Uluṟu and Kata Tjuṯa

The magnificently sculptured rock shapes of Uluṟu and Kata Tjuṯa are the product of millions of years of weathering and erosion. Indeed, some originated from geological events that occurred hundreds of millions of years ago.

The origin of the fissures along which the major valleys of Kata Tjuṯa have formed may lie in events which took place during the Alice Springs Orogeny (about 400 million years ago). Joints (fissures) formed as movement of the Mount Currie Conglomerate caused it to fracture almost at right angles to the bedding. Joints differ from faults in that there is little movement of the rocks on either side. Rain water seeping into the joints removed the rock on the sides of the joints grain by grain, over millions of years, as it flowed along and formed the canyons seen today.

Smaller vertical joints, confined to individual beds in Kata Tjuṯa, have been partly eroded and enlarged to form small caves which are like slit windows in an ancient castle.

As the surrounding and overlying rocks were eroded away, the release of confining pressure produced curved fractures called topographic joints. These can be seen at several places at Uluṟu (e.g. Ngaltawaddi, the Kangaroo Tail) and near the summit of some of the domes of Kata Tjuṯa. The layer of rock above the fracture eventually weathers away, and a new fracture forms. Weathering in Kata Tjuṯa has, to some extent, been controlled by topographic joints which have helped to produce the characteristic rounded forms.

Many of the smooth rock surfaces at Kata Tjuṯa cut cleanly across boulders contained in the conglomerate. The boulders originally protruded from the surface; their exposed parts expanded and contracted as temperatures fluctuated, whereas beneath the surface they were protected, and the resulting tensions split the boulders. The exposed and fragmented pieces were eventually washed away, leaving the smooth surfaces.

There are valleys in Uluṟu that have formed by water eroding away slightly softer beds — often the same vertical rock strata on either side of Uluṟu. Joints have not been a major influence in shaping Uluṟu. Rain-water pouring off the rock face has formed steep valleys containing a series of potholes and, in places, beautifully sculptured pools. The alignment of the valleys is mostly parallel to the bedding. The surface of Uluṟu is covered by flakes of rock, except on the very newest surfaces exposed by rock falls. The flakes result from weathering of the rock by water, and by oxygen in the atmosphere, which chemically decay minerals in the rock. The characteristic rusty colour of the exposed surface of flakes is caused by the oxidation of iron, whereas the unoxidised rock is greyish.

Weathering has produced parallel raised ridges or ribs that outline the strata, which may be due to subtle, and so far unknown, differences in the grainsize of the arkosic sands, or to slight differences in the composition or degree of cementation of the arkose.

Some scientists have suggested that sand-blasting by sandstorms has been a major influence in the shaping of Uluṟu. However, as sand is raised only a few metres above the ground during a sandstorm, it could only affect Uluṟu near ground level. Furthermore, sand-blast erosion would have removed the flakes of rock found over most surfaces of Uluṟu and left behind a roughened or polished surface. The consensus of opinion is that sand-blasting has had little affect on the shaping of Uluṟu and Kata Tjuṯa, and in any case would have been a significant factor only in the last half a million years, when the climate has been more arid.

Caves on the western face of Uluru, forming about 30 m above the present plain level

The origin of the caves of Uluṟu is a matter of scientific debate. One theory proposes that the flaky weathering process works its way into the rock unevenly. Starting from small dimples, weathering eats into the rock to form hollows. Flaking is more intense in the inner and upper shadier part of the hollow. The hollow therefore grows backwards and upwards and becomes a cave. Another theory proposes that the caves originally developed below ancient landsurfaces. Moisture trapped in the sands just below the surface helped to chemically corrode the rock and form the caves, which are now exposed due to lowering of the surrounding surface. The origin of the closely-spaced caves high on the northeastern and western faces of Uluṟu is also a subject of debate. They form brain- or honeycomb-like groups of small caves which could have formed by the flaking process.

Topographic joints, or sheet joints, on the western face of Uluru

Photographs by Colin Totterdell

Deep gorges at Kata Tjuta are eroded along joints, or fissures, in the Mount Currie Conglomerate

The near-vertical strata of Uluru are highlighted by weathering, which results in the formation of raised ribs

Mutitjulu

Weathered, flaky surface of Uluru

Some questions answered

Why do Uluṟu and Kata Tjuṯa stand out above the sand?

This is perhaps the most commonly asked question about these huge masses of rock. We know much of the answer, but not all. Some of the reason, at least, relates to the difference in hardness and resistance to erosion between the older, Proterozoic rocks and the Uluṟu Arkose and Mount Currie Conglomerate. As can be seen in the geological cross-sections and solid geology map, both the Uluṟu Arkose and Mount Currie Conglomerate are faulted against the less resistant Proterozoic rocks along their southern margins. At Uluṟu, older rocks may be present to the northeast, and younger rocks to the southwest. Since they are less resistant rocks than the arkose, the upstanding nature of Uluṟu can be partly explained. A major puzzle is the northwestern side, where we believe that Uluṟu Arkose does extend out under the sand. Perhaps the arkose here is strongly jointed, and therefore more susceptible to weathering and erosion. Conversely, the lack of major joints in Uluṟu itself means that it is very resistant to weathering and erosion, and has worn down more slowly than the jointed arkose.

Like Uluṟu, Kata Tjuṯa is faulted against less resistant, more fractured rocks along its southern margin, and possibly to the southwest as well. It has been argued that Kata Tjuṯa, Uluṟu, and Mount Conner all lie in a zone which was part of a watershed, i.e. that at some stage between 300 million years ago and the present day the whole zone was of higher relief than the surrounding country, and that streams flowed away, presumably to the north and the south, eventually leaving isolated blocks of more resistant rock. However, conglomerate is present beneath the sand for over 15 km to the northeast, and for more than 40 km to the northwest, yet only a small portion of it has been sufficiently resistant to weathering and erosion to survive as high hills. Major joints are present at Kata Tjuṯa but they are widely spaced. They have clearly played an important role in the formation of the large dome-shaped massifs. Near the faulted southern margin of Kata Tjuṯa the joints are closer-spaced, and the domes are much smaller. The unanswered question is why conglomerate to the northeast and northwest do not form high domes.

How far beneath the sand do Uluṟu and Kata Tjuṯa extend?

Uluṟu is not a giant boulder, as one popular view would have us believe. The huge vertical 'slab' of rock, of which Uluṟu is but the exposed tip, extends far below the surrounding plain, probably for several kilometres, as shown in the cross sections, as an integral part of the earth's crust.

The exact depth to which the folded layer of Uluṟu Arkose extends cannot be estimated. On the other hand, we have a very good estimate of the depth to which conglomerate extends beneath Kata Tjuṯa. Since the conglomerate layers present at the summit of Mount Olga are near the top of a 6 km thick layer of Mount Currie Conglomerate, we can predict that conglomerate should extend 6 km beneath the summit, or about 5.5 km beneath the plain in the vicinity of Mount Olga and Olga Gorge, and perhaps a little over 5 km at the Valley of the Winds.

A tour around Uluṟu

Many interesting geological features can be seen on a walk around the base of Uluṟu or in a climb to the top. These are numbered on the accompanying map and described below.

1. Base of 'The Climb'

This offers the first close-up glimpse of what Uluṟu is made of. Notice that the arkose contains faint layers roughly parallel to the direction of the climb. This is bedding (see glossary), and is beautifully displayed at Stop 2. The arkose consists of grains of sand and pebbles up to about 10 mm across; the mixture of these gives the rock a speckly appearance. For descriptions of features higher on The Climb see Stop 8.

Air photo NT Dept. of Lands & Housing

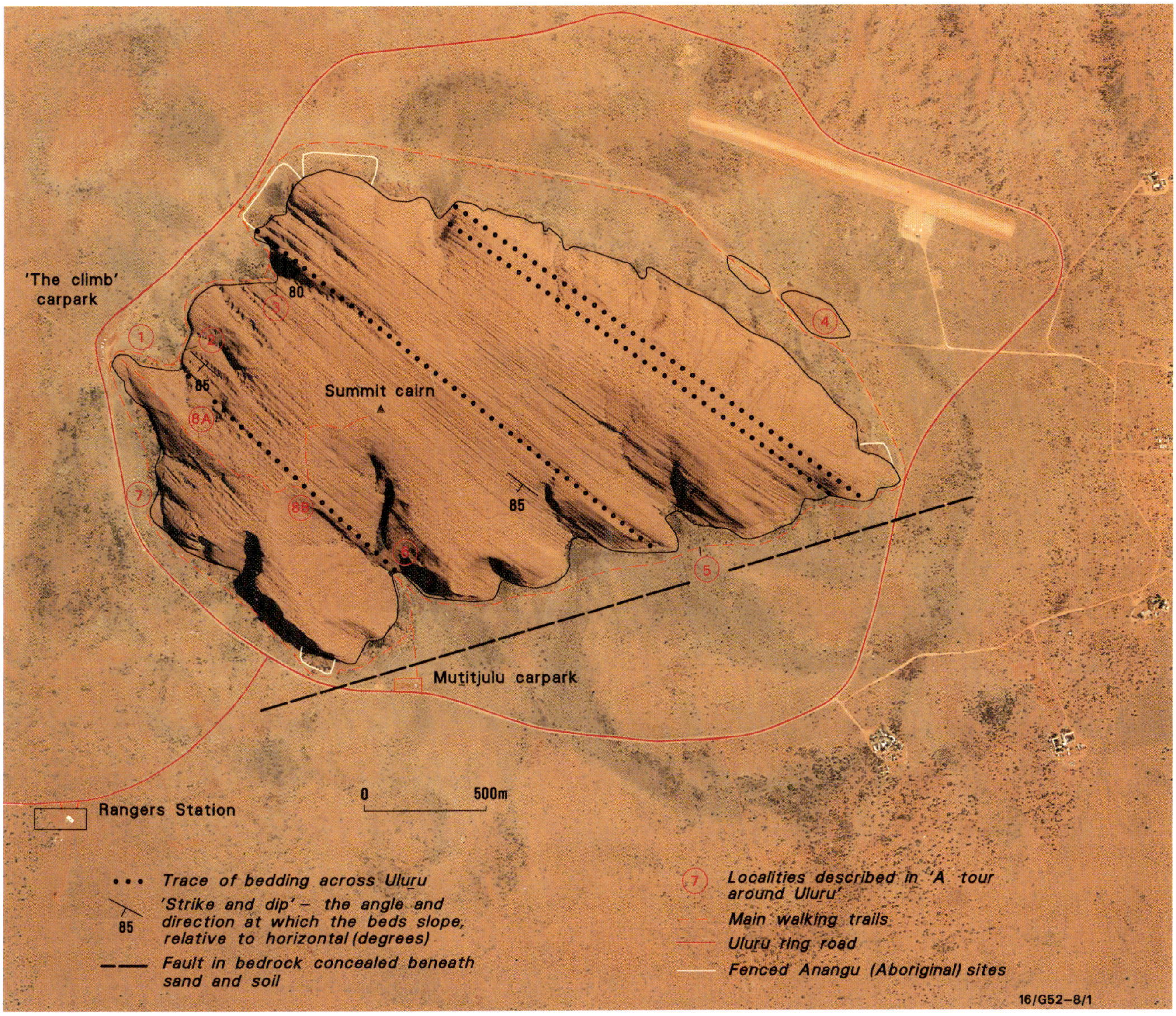

Geological map of Uluṟu superimposed on an air photograph taken from an altitude of 7 600 m

2. Malaku Wilytja (Wave Cave)

The internal structure and mineral composition of the Uluṟu Arkose are magnificently displayed here. Many geomorphic features of Uluṟu can also be seen here, making this stop the most important in the whole tour.

Bedding

The bedding, or sedimentary layering, of the Uluṟu Arkose is expressed in dark-grey or greenish-grey layers running from the roof of the cave to the floor, and sloping at about 85 degrees to the right as we look into the cave. In detail, these layers are an intricate network of curved and straight laminae, and reflect the sand and gravel layers that formed successive surfaces of the alluvial fan of which Uluṟu was once part. The layers were originally nearly horizontal. The older layers are to the left, and were progressively covered by younger layers. Bedding can be seen in most of the caves around the base of Uluṟu, always dipping in the same direction. You need to envisage that Uluṟu is but a small part of a succession of layers, at least 2.4 kilometres thick, which has been folded and faulted, ending up in a vertical attitude. The direction of rotation of the strata must have been clockwise about a horizontal axis running into the back of the cave (relative to a person standing looking into the cave). Much of this rock succession has been eroded away, but the preserved part of it extends below the ground surface to a depth of several kilometres before it bends back up towards the surface again (see cross-sections). Notice that the bedding is not simple parallel layering, but an intricate series of small curved segments, each a few centimetres to perhaps 20 cm deep, by 1 m or so in length. The curves are all concave to the right —i.e. in the direction of the younger layers. As you follow the layers up towards the cave roof they become less curved, more nearly parallel. This arrangement of layers is known as cross bedding, and helps to show in which direction the layers get younger, and from which direction the sediment came.

Curved back wall and roof of Malaku Wilytja, with ribs of resistant arkose and thick iron-oxide deposits on the overhanging roof

Composition

Malaku Wilytja is also the best place to see the composition of the Uluṟu Arkose. The fresh rock, where it has not been altered by moisture and the atmosphere, is light to dark-grey, greenish or pinkish-grey. The larger white grains, commonly standing out from the wall of the cave, are feldspar, while the slightly more translucent grains are quartz. The dark colour of some of the layers results from concentrations of iron or titanium oxide grains, probably derived from basalts in the mountains from which the sediment was derived.

Shell-like projections from the roof of Malaku Wilytja are composed of arkose that is resistant to weathering

Colin Totterdell

Weathering features and the origin of Malaku Wilytja

Parts of the lower back wall of Malaku Wilytja have a discontinuous coating of reddish iron oxide, a product of weathering of arkose, whereas most of the upper walls reveal fresh rock. The coating appears to be flaking away, suggesting that weathering in this cave is more active at present than at some time in the recent past. The lack of coatings on the upper walls suggests that flaking and enlargement of hollows or dimples above the ground surface, as described earlier, provides a plausible mechanism for the formation of this cave. The overhanging cave roof is heavily coated with iron-oxide, indicating very slow weathering and an absence of flaking. Several knobs and ribs project from the back wall and roof of Malaku Wilytja; some are coated with bird droppings. Although they look like giant bones or shells they are made up simply of arkose which is slightly harder than that surrounding them. They are likely to have formed as a result of movement of water through the rock, depositing somewhat more resistant cement from solution.

Bedding at Malaku Wilytja

3. Kantju Gorge

A vertical fracture in one of the walls of Kantju Gorge extends from the base to the top, where the gorge has cut to its deepest point. To some extent, it may have influenced the formation of the gorge.

4. Taputji (Little Ayers Rock) and rock pavements

The oldest rocks at Uluṟu are at Taputji. We can tell this from the direction of the cross bedding, i.e. that the oldest beds are to the northeast. Two hundred metres northwest of Taputji ('before' Taputji, if you are walking from Kantju Gorge) the path crosses a flat pavement of Uluṟu Arkose. The rock appears identical to the remainder of the arkose, and is just as hard, yet it has been eroded to ground level. It leaves us wondering whether the arkose extends even further out to the northeast beneath the sand — i.e. that Taputji is not the oldest part of the arkose layer, but merely the oldest exposed part. The many large blocks of weathered arkose littering the surface at Taputji are the remains of a sheet of arkose which split off from the underlying rock during development of topographic joints.

5. The faulted southern face of Uluṟu

Water bores drilled within a kilometre of the southern face of Uluṟu penetrated sand, underlain by Proterozoic siltstone and limestone or dolostone but no Uluṟu Arkose. None of these rocks are exposed, but they do confirm that this relatively straight side of Uluṟu is the result of a fault. A striking feature of the southern face is the series of caves whose bases all lie about 30 m above plain level, and whose tops are up to 65 m above plain level. These may have formed when the level of the surrounding plain was at that height. Another striking feature of the southern face is the oversteepening of the lowermost 3 to 5 m. This is known as a 'flared slope', formed by weathering at and just below ground level, implying that there has been a lowering of the level of sand around Uluṟu by 3 to 5 m. This slope lacks significant red weathered flakes of arkose, suggesting that it developed relatively recently.

Composition of the rock

Very coarse arkose, some of it conglomeratic (i.e. containing grains more than 2 mm across), is well exposed in the smooth basal portion of the flared slope. The bedding, unlike that in Malaku Wilytja, commonly consists of straight, streaky patches of conglomerate within sandstone, rather than in cross-beds, and in a few cases there are beds of conglomerate up to 0.5 m thick. These are poorly sorted, i.e. they contain a jumble of pebbles of various sizes up to about 30 mm across set in sand and silt grains of various sizes. Beds such as these have been seen only in this part of Uluṟu, and in a layer at 8B on the path to the summit. They are thought to have been deposited by 'debris flows'. Debris flows, consisting of a slurry of mud, sand and gravel lubricated by water (and called 'mud flows' if they contain a high proportion of mud), spread out across alluvial-fan surfaces when flash floods carry so much sediment that they form a slurry rather than a watery stream. Mud flows and debris flows were common in and prior to the early Palaeozoic, before the advent of land plants with their anchoring effect on rocks and sediment.

A distinct change in slope at the southeastern extremity of Uluṟu, at the same elevation as the line of caves developed on the southern face

Colin Totterdell

6. Muṯitjulu (Maggie Springs)

Water flowing off the top and slopes of Uluṟu in this area has produced a spectacular series of beautifully sculptured plunge pools and excavated a depression at the base that contains a permament pool of water.

7. The western face of Uluru

The steep western face of Uluru, visible from the road after you have passed the Rangers' Station, has many weathering features seen elsewhere. Topographic jointing is present high up on the rock between the T-junction on the Uluru ring road and The Climb. Groups of small caves, with a honeycomb-like appearance, are present nearby. A large cave in the lower slopes is at a similar elevation to those on the southern face, and presumably formed in the same way.

The Climb

Colin Totterdell

8. Climbing Uluru

To the left of the trail on the first 20 m of The Climb is a rock ledge a metre or so high consisting of large blocks of arkose. These are the remnants of a sheet of rock which has split away during formation of topographic joints parallel to the surface (see p.15). From this vantage point it is possible to see, 500 m to the northwest, an isolated outcrop of Uluru Arkose, protruding from the sand plain. The arkose is identical to that at The Climb, suggesting that much of the plain northwest of Uluru is underlain by Uluru Arkose at quite shallow depths. After you have climbed to the top of the first chained section, look for scars left by lightning. One such scar can be traced along the surface of the rock between the last pole of the first chainway and the first pole of the second chainway (8A). Lightning strikes Uluru during thunderstorms and produces scars on the surface in many places. When a lightning bolt hits heat is produced during electrical discharge along the surface of the arkose. It is sufficient to boil water trapped under the already weathered flakes of rock, and explosively removes them from the surface and exposes fresher, greyer rock underneath. Climbing the rock is not recommended during thundery weather!

Lightning scars near The Climb

After an initially steep part, The Climb winds across the top of Uluru, passing over many steep ribs, up to several metres high, which result from the comparatively rapid weathering and erosion of some arkose beds, leaving the more resistant beds upstanding. There are rarely any discernible differences in appearance between the arkose in the recessive and that in the resistant zones, but we believe that there must be subtle differences in grain composition or cement. At 8B, on top of one of these ribs, is a distinctive bed 1m thick of more pebbly arkose than elsewhere on the trail. It is thought to be another debris-flow bed, like the one described from the southern face of Uluru. Hollows of all sizes are also present on the summit of Uluru. These are presumed to form where the rock is slightly weaker, or more susceptible to attack by water and chemical action. Once water can collect, the weathering processes are accelerated, leading to a deeper or broader hollow. After rain, the water-filled hollows on the top of Uluru are a beautiful sight.

Groups of small caves high on the western face of Uluru

Panorama of Kata Tjuṯa looking north from the sand dune viewing platform

A tour around Kata Tjuṯa

Kata Tjuṯa Viewing Area

This locality provides a marvellous panoramic view of Kata Tjuṯa, and helps viewers to gain an impression of the huge thickness Mount Currie Conglomerate, of which Kata Tjuṯa is composed. The dip of the conglomerate beds is at 12 to 15 degrees to the left (west), and you can see from the sketch below that the layers on the right extend beneath those on the left and are probably at a depth of almost 2 km directly below Mount Olga. More of the conglomerate, outcropping between Kata Tjuṯa and Yulara, underlies that described above. When that is taken into account, we can estimate that around 6 km of conglomerate is present at and beneath Mount Olga.

Bedding at Kata Tjuṯa accentuated by the sun angle

Olga Gorge

1. The entrance to Olga Gorge

From the carpark the outstanding geological feature is bedding, which appears to slope west in the main gorge walls, but can be seen to actually slope obliquely from left to right and towards the viewer (i.e. roughly southwest) at about 15 degrees. The north wall of the gorge is strongly pock-marked by a series of small caves which highlight bedding, probably as a result of boulder-rich beds weathering out more easily than those containing more of the fine matrix (which are more strongly cemented). Slit-window caves low on the western end of the wall on the right, and higher up on the dome on the left, are also clearly related to a single bed.

2. Level with the ends of the domes

Over the first 100 m or so of outcrop, the conglomerate surface is covered with reddish-brown iron oxide, derived from weathering of minerals in the conglomerate. Fresh rock is visible in several small gullies, in which the abrasive action of sediment carried by running water has prevented the build-up of iron oxides. You can see that the conglomerate is composed of large (up to 0.5 m across), rather smooth, rounded clasts (see glossary) set in a matrix of pebbles and finer material. The clasts include light coloured, speckly granite, and dark bluish to greenish-grey ones of basalt. Many of the basalt clasts in turn contain speckles of light coloured minerals — these are called amygdales, and are the fillings of gas bubbles which developed in the basaltic lava as it flowed to the surface in the source regions for the conglomerate. There are also fragments of various metamorphic rocks, of which the most common are light-coloured, banded or streaked ones called gneiss. The matrix, the

Sketch of panorama of Kata Tjuṯa showing conglomerate layers projected above and below ground level . The scale AB represents 500 m of conglomerate, & demonstrates that about 1 800 m of strata are present at Kata Tjuṯa

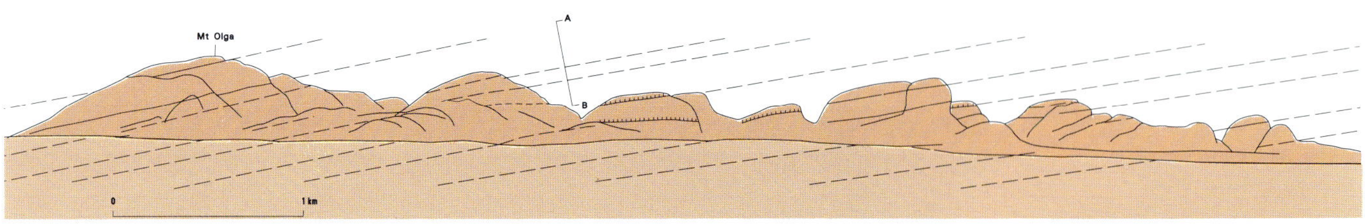

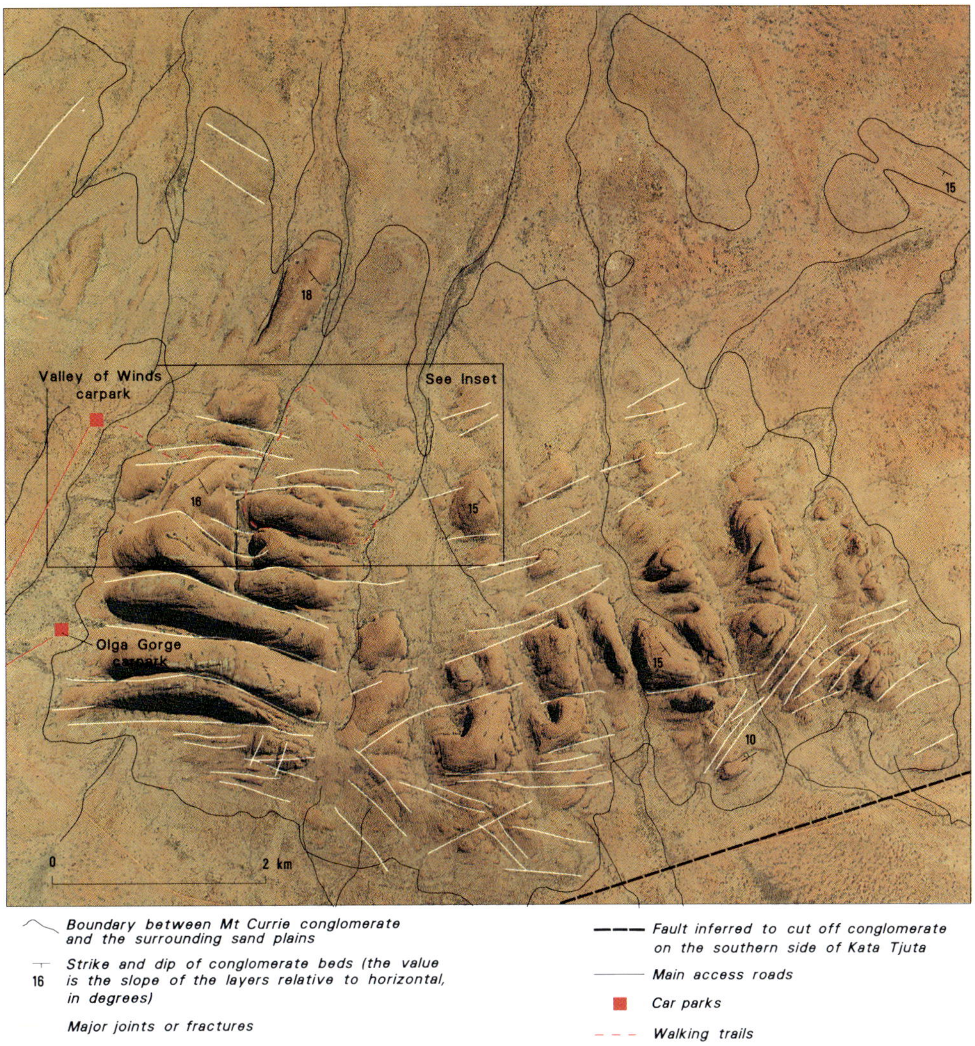

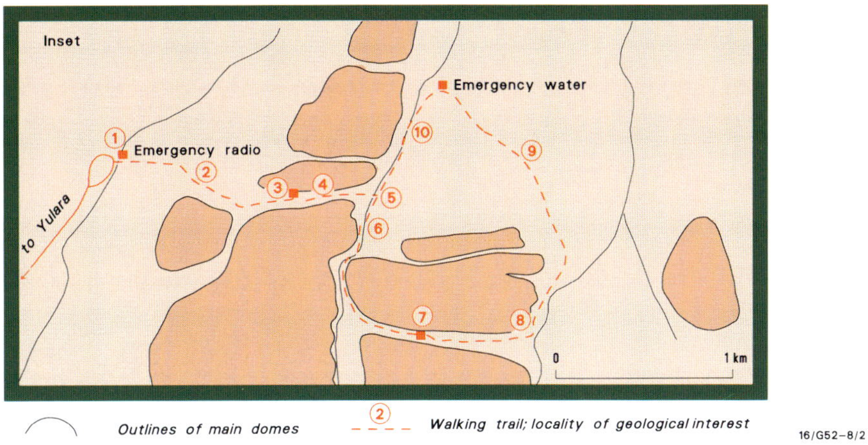

Geological sketch map of Kata Tjuṯa

natural cement which binds the clasts together, transforming what was gravel into conglomerate, is generally green. The colour results from the presence of the mineral epidote, formed from a previously fine muddy matrix under the effects of heat and pressure when the Mount Currie Conglomerate was deeply buried. Note that over most of the conglomerate surface, particularly on the lower slopes of the domes, boulders do not protrude but are fractured to form a relatively smooth surface.

The view away from the gorge to the west and southwest reveals a series of ridges on the sky line. These are the eroded remnants of mountains containing granites and metamorphic rocks similar to those present in the Mount Currie Conglomerate, but also quartzites, which are seen only in the oldest conglomerate layers. The mountain ranges from which the Mount Currie Conglomerate was derived, over 550 million years ago, (or at least their margins) lay in the same direction, but probably closer to Kata Tjuṯa.

Where the walking trail reaches its highest point and begins to descend towards the creek in the gorge, it passes over two ledges of greenish epidote-rich sandstone. Similar sandstone layers, rarely more than 1 metre thick, form wispy traces in the walls of the gorge, particularly in the lower part of the north wall.

3. Conglomerate blocks

On entering the gorge, note the jumbled piles of large blocks of conglomerate littering the gorge floor. In many instances it is possible to see fractures in the walls above, showing where these blocks originated. Most of the falls are quite old, but some are relatively recent, as shown by the lack of an iron oxide coating on the fallen blocks. (Be warned; although rare, rock falls continue to the present day!)

Northern wall of Olga Gorge. The large blocks of conglomerate have fallen from the area in shadow near the centre of the photo .

4. End of Gorge walk

Note the many thin sandstone beds in the south wall. The gorge floor is choked by debris fallen from the walls, consisting mostly of small angular rock chips, sand and dust, derived from the continual flaking of material. The base of the northern wall, which is accessible here, consists of a very smooth pavement in which the splitting of clasts at the surface can be clearly seen. If you gaze upwards you will see sheets of rock as thin as 1 cm splitting off from the rock surface; it is fragments of these sheets which have built up on the floor of the gorge.

Valley of the Winds walk

1. From the carpark

From the carpark, bedding, dipping at around 16˚ to the southwest, is clearly visible in the large dome to the east. If you turn and look west you will see low hills composed of the same conglomerate but eroded to a much lower level. Similar conglomerate extends to the west in a series of discontinuous outcrops for over 20 km.

2. Conglomerate pavements

300 m from the carpark, on the left, are the first conglomerate outcrops. They are mostly rubble covered, with whole boulders partly eroded out from the matrix. There are also some patches where the exposed parts of boulders have split away, leaving almost flat pavements.

A 1.5 m granite boulder in the Mount Currie Conglomerate. The darker coloured boulders are basalt

3. Sandstone beds

Bedding is clearly visible in the dome to the south of the first lookout. It is highlighted by streaky, lighter coloured sandstone beds. Near the base of the dome an area of dark greyish-green rock, with a very smooth face, is a site from which a large slab of conglomerate has split off, cutting through boulders and matrix alike, exposing the fresher, greener rock. Individual boulders show up more clearly in such places. Such splitting is described above as topographic jointing. The trail turns sharply right at the first lookout, and descends for 40 m over a relatively smooth, undulating pavement of sandstone. This sandstone bed is one of the largest seen at Kata Tjuṯa; Most others are lens-shaped bodies up to 1 m thick, and a few metres long, within the conglomerate. Note the greenish tinge due to the mineral epidote, and the sheet-like flaking of its surface. Many of the sandstone lenses are cross-bedded, enabling us to measure the direction in which the sand was being carried at the time of deposition (and indirectly, to guess where it may have come from).

4. Topographic jointing

About half way from the first lookout to Stop 5 a 2 to 3 m high wall of conglomerate on the left is separated from the conglomerate beneath it by a large fracture. This is clearly another topographic joint, caused by unloading near the surface. Note a large (nearly 1 m) granite boulder which appears to be on the verge of crashing onto the path (This boulder weighs over half a tonne, is already partly dislodged, and is potentially dangerous. Do not touch it!). Further along, the trail winds past large blocks of conglomerate which have already tumbled down the slopes.

5. At the footbridge

From the footbridge a series of thin lines of spinifex are visible near the base of the dome you have just passed (to the southwest). The plants are taking advantage of the extra moisture and soil developed in a series of minor joints.

Lenses of sandstone at bottom and top of photo separated by pebble and cobble conglomerate

6. At the creek crossing

At the creek crossing the conglomerate and its constituent boulders are well exposed, and 10 m south of the crossing the conglomerate matrix is unusually well displayed in a large fallen block. In this block, large rounded boulders are cemented together by a 'matrix' consisting of much smaller clasts, ranging right down to smooth pebbles and sand grains. These in turn are bound by a dark greyish-green material, composed largely of epidote. This was originally fine silt and mud, probably derived from the rapid erosion of a basalt-rich terrain.

7. Valley of the Winds Lookout

Like Olga Gorge, the gorge at the Valley of the Winds Lookout is eroded out along a major vertical fracture or joint in the conglomerate. This joint may have formed as much as 400 million years ago, but the present erosion of it has presumably occurred during the Cainozoic, over the last 50-60 million years! The walls are very steep, and pock-marked by weathering. In overhanging parts of the southern wall, streaks of iron oxide, and perhaps lichen, have built up as a result of the seepage of water from above. The large domes to the east are all composed of conglomerate which is indistinguishable from that at the main lookout. Because the bedding in the domes slopes to the west-southwest, the layers of conglomerate in the distant domes are present at about 1 km depth beneath you at the lookout.

Mount Currie Conglomerate at Kata Tjuṯa

8. Bedding

Individual beds in the isolated dome to the east-northeast seem discontinuous, and many are at slight angles to one another. A thin spoon-shaped layer of sandstone is also faintly visible. These are all features of bedding which reflect the way in which gravel built up on the Kata Tjuṯa alluvial fan, probably in shallow channels, over 550 million years ago.

9. Rock surfaces

On the left, 150 m south of the Emergency Water Point, the brown oxidised coating on the conglomerate is absent because of the continual action of modern sediment being carried down the creek. The different types of boulders in the conglomerate can be clearly seen. About 100 m further south, on the gentle slope left of the track, is an area of rock covered in a white crust. This is lime, derived from the weathering of minerals in the conglomerate, and deposited from groundwater seepage.

Desert oaks

Colin Totterdell

Glossary

alluvial fan. A gently sloping plain at the foot of a mountain range, built up of sand and gravel deposited from streams issuing from narrow mountain valleys after heavy rains. The shape of the plain commonly resembles that of an open fan, hence the name.

arkose. Sandstone containing abundant grains of feldspar, commonly pink or reddish to pale grey or buff, usually derived from the rapid erosion of granitic rocks.

basin. See sedimentary basin.

bed/bedding. Layers in sediments or sedimentary rocks, reflecting differences in grain-size, composition or colour of the constituent grains.

calcrete. Lime cement or crust, formed just under or at the land surface, generally as part of soil layers.

clast. A grain or fragment making up part of a sedimentary rock, but usually restricted to the large fragments making up conglomerates. By convention, clasts are described as granules if their longest dimension measures 2 to 4 mm, pebbles if between 4 mm and 64 mm, cobbles if between 64 and 256 mm, and boulders if over 256 mm.

conglomerate. A sedimentary rock containing fragments greater than 2 mm across (granules, pebbles, cobbles and boulders).

cross-bedding. A distinctive and common form of bedding in sedimentary rocks, resulting from sand and other material being laid down in ripples or dunes, either under water or by the wind. The structure is often seen in sandstones as straight or curved layers at angles of up to 35° to the major bedding surfaces.

dolomite/dolostone. Dolomite is a mineral made up of equal amounts of calcium and magnesium ions bonded together with carbonate ions; $CaMg(CO_3)_2$. Dolostone is a sedimentary rock composed largely of dolomite.

epidote. A yellowish-green silicate mineral containing calcium, aluminium, and iron. It is most common in metamorphic rocks.

fault. A fracture or zone of fractures along which the rocks have been displaced.

igneous rock. A rock resulting from the solidifying of magma. If the magma solidifies deep beneath the earth's surface, coarse-grained rocks such as granite are formed. If the magma flows to the surface it generally forms fine-grained volcanic rocks, e.g. basalt, andesite, or rhyolite.

limestone. A rock consisting of grains, or fragments of animal shells, of calcium carbonate ($CaCO_3$)

magma. Molten rock formed within the earth's crust or upper mantle. Magma which is erupted onto the earth's surface becomes lava.

metamorphism/metamorphic rocks. Metamorphism is a geological process in which the minerals in solid rocks change form owing to the heat and pressure to which they are subjected deep below the earth's surface. Metamorphic rocks are rocks formed through such processes; they may form from igneous rocks, sedimentary rocks, or older metamorphic rocks.

mudstone. A sedimentary rock composed originally of 'mud', a mixture of clay and silt. It is similar to, but lacks the flaky nature of, shale.

orogeny/orogenesis. The folding, faulting, and (usually) uplifting, over an extended period, of a part of the earth's crust (literally: 'mountain building').

rhyolite. A rock formed from silica-rich lava. It commonly contains crystals of quartz and feldspar set in a fine 'groundmass' of glassy material.